M000287053

WORKBOOK FOR THE FUNDAMENTALS OF SOUND SCIENCE

FIRST EDITION

WRITTEN AND EDITED BY Elena Borovitskaya

Temple University

Bassim Hamadeh, CEO and Publisher
Kassie Graves, Director of Acquisitions
Jamie Giganti, Senior Managing Editor
Jess Estrella, Senior Graphic Designer
John Remington, Senior Field Acquisitions Editor
Kaela Maritn, Project Editor
Natalie Lakosil, Senior Licensing Manager
Rachel Singer, Associate Editor
Kat Ragudos, Interior Designer

Printed in the United States of America

ISBN: 978-1-5165-0305-6 (pbk) / 978-1-5165-0306-3 (br) / 978-1-5165-0307-0 (pf)

Contents

SIMPLE HARMONIC MOTION AND RESONANCE

Simple Harmonic Motion

1. Give examples of simple harmonic motion.

2. Based on what you have learned in class, what are the two main elements of a system demonstrating simple harmonic motion?

3. In the axis below, schematically draw the dependence of the displacement of mass on time.

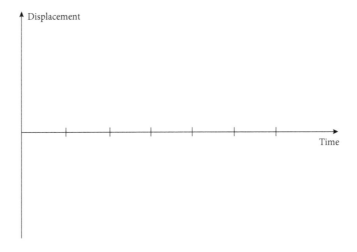

Figure 1.1

3.1 On your graph, mark a period of oscillations, T. What units do we use to measure period?

3.2 On your graph, mark the amplitude of oscillations, A. What units do we use to measure the amplitude of oscillations of mass on a spring?

4. Discuss frequency of oscillations, f.

4.1 Give the short definition of frequency:

4.2 What are units of frequency?

4.3 If the frequency of oscillations of a system is 100 Hz, what is its period?

4.4 If the frequency of oscillations of a system is 100 Hz, how many whole vibrations per second does this system make?

4.5 What is the formula connecting period and frequency of oscillations?

5. Use the following table to understand the influence of mass and the stiffness of a spring on the frequency and period of oscillations.

If _____ increases	Period is	Frequency is
mass		
stiffness		
amplitude		

6. Using the results of the table above, answer the following questions:

6.1 Increasing **what** will increase the frequency of oscillations?
    a.   Mass
    b.   Stiffness
    c.   Amplitude
    d.   Stiffness and amplitude
    e.   Mass and amplitude.

6.2 We perform two experiments with the **same** mass and spring. Amplitude of oscillations in experiment #2 is twice as great as the amplitude of oscillations in experiment #1. Which of the two experiments will show greater frequency of oscillations?
    a.   Experiment 1
    b.   Experiment 2
    c.   They both will demonstrate the same frequency.
    d.   Impossible to answer from the information given.

7. How does amplitude affect the frequency of oscillations? (Describe in your own words).

Resonance

8.1 In the axis below, schematically draw several periods of oscillations of mass on spring **without** additional force.

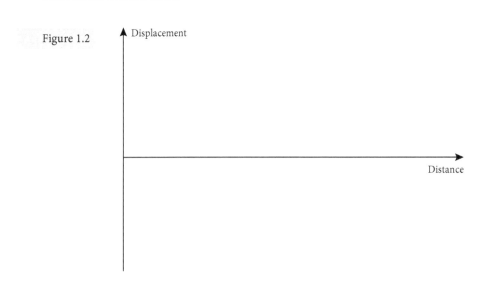

Figure 1.2

8.2 Mark with the symbol ° the points where you should apply force to reach resonance. What is the period of these points? What is the frequency?

8.3 Mark with the symbol * to point out what will also produce resonance. What is the period of these points? What is their frequency?

8.4 The forces of which periods, besides the natural one, will produce resonance in an oscillating system? To what frequencies do these periods correspond?

8.5 Now that you are familiar with the concept of resonance, explain why soldiers should break step while crossing a bridge.

8.6 To break a wineglass using sound, the sound should be
   a.   As loud as possible
   b.   As close as possible to the natural frequency of glass
   c.   As soft as possible
   d.   All of the above.

Conclusions

1. Convert the terms of physics into terms of music:
   a.   Frequency = _____
   b.   Amplitude is related to _____

2. Instead of a mass-spring system, consider the following for a guitar string:
   a.   Stiffness = _____
   b.   Mass = _____

3. Give a definition of resonance:

WAVES: BASICS

Waves

---

1. In your own words, formulate the difference between wave and vibration.

2. In the axes below, draw the dependence of displacement on time at a given position.

Figure 2.1

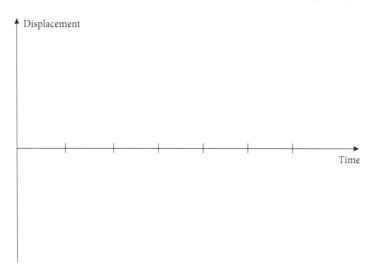

2.1 Mark a period of a wave on this graph. What units do we use to measure period?

3. In the axes below, schematically draw the dependence of displacement on a position at a given moment in time.

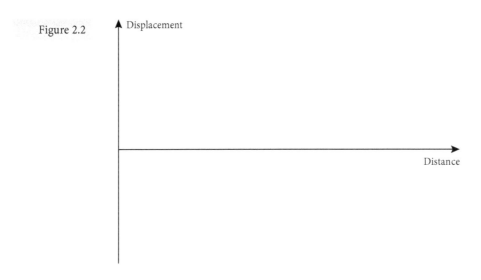

Figure 2.2

3.1 Mark the wavelength on this graph. What units do we use to measure wavelength?

3.2 Is the wave shown above transversal or longitudinal?

3.3 Give examples of transversal waves.

4. Schematically draw the snapshot of a longitudinal wave on a Slinky.

4.1 In the picture, show the wavelength of the longitudinal wave.

4.2 If your longitudinal wave travels from the left of your page to the right, in what direction will the particle of the wave oscillate?

a. Up and down
b. Left and right
c. In any direction.

4.3 Name some examples of a longitudinal wave.

4.4 If the wavelength of a longitudinal wave is 6 m, what is the distance between compression and the nearest expansion (rarefaction)?
a. 6 m
b. 3 m
c. 2 m
d. 12 m.

Sound Waves

5.1 From the list below, **what** waves may travel through a vacuum?
a. Sound and light
b. Sound only
c. Light only
d. Neither sound nor light.

5.2 Write a formula demonstrating the dependence on temperature of the speed of sound in the air.

5.3 What is the speed of sound right now (approximately) in our classroom?

5.4 How long will it take for sound to travel from the blackboard to the back wall of our classroom (approximately)?

5.5 How fast does sound travel in water? (You may use the Internet to find the data for this.)

a. Sound in water travels faster than in the air
b. Sound in water travels slower than in the air
c. Sound in water travels with the same speed as in the air
d. Sound in water does not travel at all.

6. What relationship connects the frequency, wavelength, and speed of a wave?

6.1 The longest audible for human ear sound wavelength is 20 m. To what frequency does it correspond at a temperature of 13°C?

6.2 Does the frequency in the question above correspond to bass or treble?

6.3 The shortest audible for human ear sound wavelength is 2 cm. To what frequency does it correspond at a temperature of 13°C?

6.4 Does the frequency in the question above correspond to bass or treble?

7. If the wavelength of wave A in the air is 4m, what is the wavelength of wave B, whose frequency is twice the frequency of wave A?
a. 4 m
b. 8 m
c. 2 m
d. 6 m.

8. Two waves of frequencies, 220 Hz and 440 Hz, travel in our room. We can predict that
a. The wave of 440 Hz travels faster than the wave of 220 Hz
b. The wave of 440 Hz travels slower than the wave of 220 Hz

c.   These two waves have the same speed
d.   The wave of frequency of 440 Hz can be slower or faster than the wave of 220 Hz; it depends on the particular musical instrument.

Conclusions

1. The speed of sound in the air depends on (select all that apply)
   a.   Loudness
   b.   Pitch
   c.   Wavelength
   d.   Temperature
   e.   All of the above.

2. When a sound wave travels from east to west, the particles in this wave vibrate in
   a.   An east-west direction
   b.   A north-south direction
   c.   A vertical direction
   d.   All possible directions.

FAMILIES OF MUSICAL INSTRUMENTS: BASICS, INFLUENCE OF SOURCE SIZE, AND OCTAVE

Basics

1. What families of musical instruments can you name?

2. To which family of musical instrument does each of the following belong:

Cymbal _____

Guitar _____

Tuning fork (it is not a musical instrument, but …) _____

Glockenspiel _____

Piano _____

Violin _____

Celesta _____

Flute _____

Clarinet _____

3. In your own words, describe what height of pitch you expect from small and large musical instruments.

4. What physical characteristics of a wave increases with the pitch going down?
   a.   Frequency
   b.   Wavelength
   c.   Speed
   d.   All of the above.

5. Consider two sound pipes, which are either demonstrated by your instructor or in your possession.

   5.1 What is the length of these pipes while both ends are open?

   Longer pipe _____ (m)

   Shorter pipe _____ (m)

   5.2 Read the markings on the pipes. In your book (or on the Internet), find the frequencies that correspond to these notes.

   Longer pipe note _____ frequency _____ (Hz)

   Shorter pipe note _____ frequency _____ (Hz)

5.3 Estimate the temperature in our classroom and calculate the speed of sound in our room: _____ (m/s)

5.4 Now, calculate the wavelength which corresponds to this frequency:

Longer pipe _____ (m)

Shorter pipe _____ (m)

5.5 Calculate the approximate ratio of wavelength from the previous question to the length of a pipe from question 5.1. What is this ratio?

Longer pipe _____

Shorter pipe _____

5.6 To conclude, to what fraction of wavelength do these pipes correspond:

Longer pipe _____

Shorter pipe _____

5.7 Is this ratio the same for both pipes?

6. Listen to two sounds created by hitting two bars of a glockenspiel.

6.1 What are the notes and corresponding frequencies of these sounds?

Longer bar note _____ frequency _____ (Hz)

Shorter bar note _____ frequency _____ (Hz)

6.2 What is the length of these bars?

Longer bar _____ (cm)

Short bar _____ (cm)

6.3 What is the ratio between the lengths of the longer and shorter bars? (Use the results of the previous question.)

6.4 What are the wavelengths created by these bars (use the speed of sound, which you calculated in part 5.3):

Longer bar _____ (m)

Shorter bar _____ (m)

6.5 What is the ratio between the wavelengths of the longer and shorter bars? (Use the results from the previous question.)

6.6 In your own words, describe the relationship between the lengths of pipes and their pitch and the lengths of bars and their pitch. Do you see the difference in these relationships?

Octaves

7.1 Give the definition of an octave.

7.2 Listen to a note with the frequency of 440 Hz. What are the frequencies of the note

One octave lower _____

One octave higher _____

Two octaves lower _____

Two octaves higher _____

Three octaves lower _____

Three octaves higher _____

7.3 Listen to the pipes you considered in questions 5.1–5.7, but with one end closed. What happens to a pitch when you close one end?

7.4 What frequency corresponds to a pitch that is produced by the pipes now?

Longer pipe _____ (Hz)

Shorter pipe _____ (Hz)

7.5 What is the wavelength of sound known for each pipe?

Longer pipe _____ (Hz)

Shorter pipe _____ (Hz)

7.6 What is the ratio of the length of the pipe to the wavelength of sound now?

Longer pipe _____

Shorter pipe _____

Conclusions

1. A sound three octaves above 200 Hz will have a frequency of:
   - a.   400 Hz
   - b.   600 Hz
   - c.   800 Hz
   - d.   1000 Hz
   - e.   1200 Hz
   - f.   1600 Hz.

2. If you cut off a piece of a cylindrical pipe, its pitch will:
   - a.   Go up
   - b.   Go down
   - c.   Stay the same
   - d.   Impossible to predict.

PHYSICAL PROPERTIES OF SOUND: INTERFERENCE, BEATS, AND THE DOPPLER EFFECT

1. Check your understanding of the phenomenon of interference.

1.1 What is interference, and when does it occur? Describe this in your own words.

1.2 A listener is positioned at a distance of 6m from one loudspeaker and 4.5m from another one.

1.2.1 What is the difference in the path traveled by a wave from the first and second loudspeakers?

1.2.2 If we want to know what wavelengths experience **constructive** interference, to what value should this difference be equal? (Recall the condition of **constructive** interference.)

1.2.3 What wavelength corresponds to constructive interference in the given example?

      Wavelength #1 _____

      Wavelength #2 _____

      Wavelength #3 _____

1.2.4 Estimate the temperature in our classroom. What is the speed of sound in our classroom now?

1.2.5 Using the relationship between frequency, wavelength, and speed of a wave, calculate frequencies that correspond to the wavelength in question 1.2.4.

      Frequency #1 _____

      Frequency #2 _____

      Frequency #3 _____

1.2.6 If we want to know what wavelengths experience **destructive** interference, to what value should this difference be equal? (Recall the condition of **destructive** interference.)

1.2.7 What is the first wavelength which corresponds to destructive interference in the example given?

      Wavelength #1 _____

1.2.8 Using the relationship between frequency, wavelength, and speed of a wave, calculate the frequency which corresponds to the wavelength in question 1.2.8.

Frequency #1 _____

1.2.9 Draw a schematic sketch corresponding to the path of waves for the first wavelength experiencing constructive interference.

1.2.10 Draw a schematic sketch corresponding to the path of waves for the first wavelength experiencing destructive interference.

2. Check your understanding of the concept of beats.

2.1 In your own words, define beats.

2.2 Listen to two metallic bars hit together. How would you describe what you hear?

2.3 Listen to two metallic bars hit together: one sounds at 440 Hz, the other at 439 Hz. What is the period of beats which you hear, and what is the frequency of the beats? (To measure the period of beats, use your wristwatch or the stopwatch from your cell phone.)

Period _____ (s)

Frequency _____ (Hz).

2.4 Listen to two metallic bars hit together: one sounds at 440 Hz, the other at 435 Hz. What is the period of beats you hear, and what is the frequency of the beats?

Period _____ (s)

Frequency _____ (Hz)

2.5 Based on these two experiments, what relationship between the frequencies of the hit bars and the frequency of the beats could you present?

3. Doppler effect

3.1 Describe the Doppler effect in your own words.

3.2 Let's play ball!

3.2.1 What change of pitch do you hear when the Doppler buzzer approaches you (higher, lower, or the same as at rest)?

3.2.2 How does the wavelength change when the Doppler buzzer approaches you (becomes longer, shorter, or stays the same as at rest)?

3.2.3 What pitch do you hear when the Doppler buzzer travels away from you (higher, lower, or the same as at rest)?

3.2.4 How does the wavelength change when the Doppler buzzer moves away from you (becomes longer, shorter, or stays the same as at rest)?

3.3 What characteristics of sound do not change due to the Doppler effect when the distance between a source and a receiver changes?

    a.   Frequency

    b.   Wavelength

    c.   Speed

    d.   All of the above.

Conclusions

1. A receiver is placed 7 m from one loudspeaker and 4 m from another. What is the first wavelength to experience constructive interference in this setup?

    a.   2 m

    b.   3 m

    c.   6 m

    d.   9 m.

2. Two strings demonstrate a beat frequency of 4 Hz. What would happen to the beat frequency if we decrease the tension of a string producing a higher pitch?

    a.   Beat frequency will increase

    b.   Beat frequency will decrease

    c.   Beat frequency will stay the same

    d.   Impossible to predict.

3. You hear the whistle of a train approaching you with constant speed. The pitch you hear is:

    a.   Lower than the real one, gradually decreasing

    b.   Higher than the real one and gradually increasing

    c.   Lower than the real one, and it stays the same

    d.   Higher than the real one, and it stays the same.

MEASUREMENTS OF LOUDNESS: INTENSITY

Intensity

1.1 Give the definition of intensity:

1.2 What units do we use for intensity? _____

1.3 The table below shows a wave with different amplitudes. The intensity for one of the amplitudes is given. Fill in all other intensities.

Amplitude (m)	Intensity (W/m²)
2 m	10
4 m	
6 m	
8 m	

1.4 What is the audible range of intensities for the human ear?

1.5 What is the ratio of the highest intensity audible to the human ear to the lowest intensity?

2. Sound intensity level (SIL)

2.1 What units do we use to measure sound intensity level?

2.2 What is the audible range of the SIL for the human ear?

2.3 Listen to one tuning fork, followed by two tuning forks hit together. Do two tuning forks sound twice as loud compared to one tuning fork? Briefly describe what your feel.

2.4 What is the ratio of the intensities of two tuning forks to one tuning fork? _____

2.5 To what difference in the sound intensity level (SIL) does this ratio correspond? _____

2.6 If one tuning fork produces a sound of 63 dB, what is the sound intensity level produced by two tuning forks hit together? _____

2.7 Fill in the table below for the SILs corresponding to the given intensities.

Intensity (W/m²)	SIL (dB)	Typical for
10^{-10}		
10^{-7}		
10^{-6}		
10^{-4}		
1		

2.8 What is the typical SIL for a conversation? _____

2.9 What is a musically important range of the SIL?

3. Inverse-square law

3.1 Describe the inverse-square law in your own words:

3.2 Fill in the table below. Intensity is given for distance from a source of 2 m. What are the intensities and SILs for other distances?

Distance from a source (m)	Intensity (W/m²)	SIL (dB)
2	1	
20		
200		

Conclusions

1. An SIL level created at some position by one violin is 73 dB. What is the SIL created at this point by ten identical violins positioned at equal distances from the listener?

    a.   730 dB

    b.   83 dB

    c.   63 dB

    d.   93 dB.

2. The intensity at a distance of 5 m from a sound source is 20 W/m². What is the intensity at 10 m from a source?

    a.   10 W/m²

    b.   5 W/m²

    c.   20 W/m²

    d.   80 W/m².

PERCEPTION OF SOUND: SONES AND PHONS

1. On a function generator, set some level of amplitude and leave it. Change the frequency and listen to the loudness. Describe how loudness changes when the frequency goes higher and higher.

Sones

2.1 Give the definition of a sone:

2.2 The figure below shows the relationship between the loudness in sones and the SIL in dB for the sound of 1000 Hz. On this graph, make a schematic drawing of the relationship for a wave of 100 Hz and of 3000 Hz.

Figure 6.1

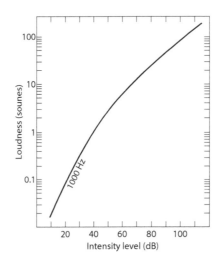

2.3 What SIL corresponds (approximately) to the loudness of ten sones for 1000 Hz? _____

Phons

3.1 Give the definition of a phon:

3.2 Using the diagram below, mark with you pencil (or pen) an equal loudness corresponding to the loudness of 70 phons.

Figure 6.2

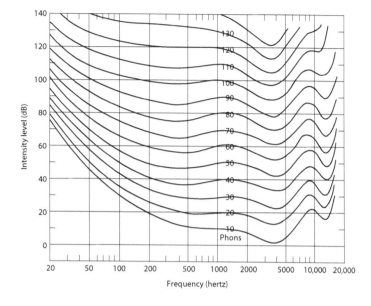

3.3 In your own words, describe the equal-loudness line.

3.4 Using the diagram above, fill in the table below. In the last column, mark sounds with numbers from 1 to 4, where 1 is the loudest sound and 4 is the least loud.

SIL (dB)	Frequency (Hz)	Loudness (phons)	Ranking
70	50		
70	200		
70	1000		
70	3000		

3.5 Using the graph below, fill in the table. Now that you have the loudness in sones, find the corresponding SIL in dB.

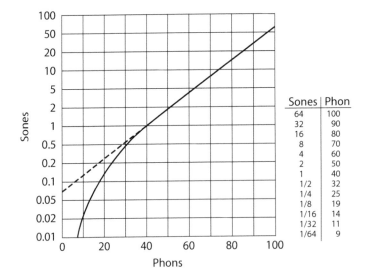

Figure 6.3

Sones	Phon
64	100
32	90
16	80
8	70
4	60
2	50
1	40
1/2	32
1/4	25
1/8	19
1/16	14
1/32	11
1/64	9

Frequency (Hz)	Loudness (sones)	SIL (dB)
100	2	
200	8	
2000	32	

3.6 Using the graphs given in this activity, fill in the following table. Now the loudness in sones is the same for all frequencies. Find the corresponding loudness in phons and the SIL that will provide this level of loudness.

Loudness (sones)	Frequency (Hz)	Loudness (phons)	SIL (dB)
4	200 Hz		
4	1000 Hz		
4	4000 Hz		

Conclusions

1. What SIL corresponds to a 500 Hz wave with a loudness of four sones?
   a. Between 60 and 70 dB
   b. Between 70 and 80 dB
   c. Between 50 and 60 dB
   d. Between 40 and 50 dB.

FOURIER ANALYSIS OF THE SIMPLEST SOUND SPECTRA

Harmonic Series

1.1 Give the definition of a harmonic series.

1.2 Give an example of a harmonic series:

Fourier Spectra: First approach

2.1 In the axes below, schematically draw three graphs: in the upper system of coordinates, draw two periods of a sinusoidal wave with a period T (frequency f). In the second system of coordinates, draw four periods of a sinusoidal wave with a period T/2 (frequency 2f). In the lowest graph, add these two functions together. Try to do it carefully.

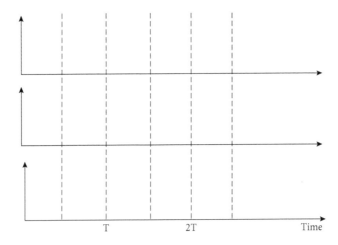

Figure 7.1

2.2 What is the frequency of repetition of your resulting wave? _____

2.3 Does this function look sinusoidal? _____

2.4 If you add another sinusoidal function of period T/3 (frequency 3f) to the sum of functions, will it change the frequency of repletion of the resulting wave? (If you have any doubts, try to draw it on a graph): _____

2.5 Formulate what result you expect if you add sinusoidal functions that **do belong** to the same harmonic series:

2.6 Formulate what result you expect if you add sinusoidal functions that **do not belong** to the same harmonic series:

Fourier Analysis: Square Wave

3.1 Fill in the table below for the **square wave**. Amplitude and intensity of the fundamentals are given. What are the amplitudes and intensities of other harmonics?

# of harmonic	Amplitude	Intensity
1	100	81
2		
3		
4		
n		

3.2 For the last position in a table, can n be any number or only an odd number?

Triangle Wave

4.1 Fill in the table below for the **triangle wave**. Amplitude and intensity of the fundamentals are given. What are the amplitudes and intensities of other harmonics?

# of harmonic	Amplitude	Intensity
1	100	81
2		
3		
4		
n		

4.2 For the last position in a table, can n be any number or only an odd number?

Sawtooth Wave

5.1 Fill in the table below for the **sawtooth wave**. Amplitude and intensity of fundamentals are given. What are the amplitudes and intensities of other harmonics?

# of harmonic	Amplitude	Intensity
1	100	81
2		
3		
4		
n		

5.2 For the last position in a table, can n be any number or only an odd number?

Modulated Sounds: Vibrato and Tremolo

6.1 In the axes below, show the typical spectrum of tremolo.

Figure 7.2

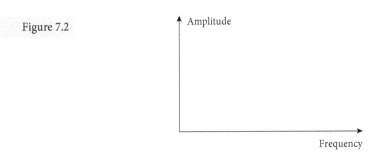

6.2 What is the result of using tremolo?
    a.   Broadening of existing lines
    b.   Appearance of additional lines in spectra
    c.   Shifting of lines toward lower frequencies
    d.   Shifting of lines toward higher frequencies.

6.3 In the axes below, show the typical spectrum of vibrato.

Figure 7.3

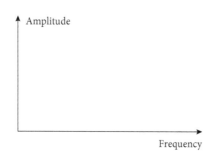

6.4 What is the result of using vibrato?
    a.   Broadening of existing lines
    b.   Appearance of additional lines in spectra
    c.   Shifting of lines toward lower frequencies
    d.   Shifting of lines toward higher frequencies.

7. Check your understanding of the material.

7.1 The figure below shows the axis of frequencies with the audible range for the human ear.

Figure 7.4

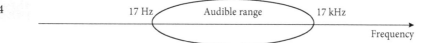

7.1.1 Schematically show the harmonics created by the **square** wave with fundamental 5 Hz. Will this wave be audible? _____

7.1.2 Schematically show the harmonics of the **sinusoidal** wave with fundamental 10 Hz. Does the **sinusoidal** wave have any fundamentals? Will it be audible? _____

7.1.3 Now, schematically show on the axis above the harmonics for the **square** wave with fundamental 20,000 Hz. Will this wave be audible? _____

7.1.4 Schematically show the harmonics of the **sinusoidal** wave with fundamental 25,000 Hz. Does the **sinusoidal** wave have any fundamentals? Will it be audible? _____

1. What is the only wave which DOES NOT have any upper harmonics?

      a.   Square wave

      b.   Triangle wave

      c.   Sawtooth wave

      d.   Sinusoidal wave.

2. Of the waves we studied, which one has both even and odd harmonics?

      a.   Square wave

      b.   Triangle wave

      c.   Sawtooth wave

      d.   Sinusoidal wave.

BASICS OF PERCUSSION INSTRUMENTS AND NORMAL MODES: CHLADNI PLATES AND BARS

Metallic Bars

1.1 Based on the demonstration, schematically draw the first three modes of a metallic bar clamped from one end. Mark the nodes (n) and antinodes (a).

1.2 Measure the length of two metallic bars on the glockenspiel with a different pitch by the octave. Length of

Longer bar _____ (cm)

Shorter bar _____ (cm)

1.3 What is the ratio of lengths? _____

1.4 What is the ratio of the frequencies of these two bars (don't forget, it's an octave!)? _____

1.5 Can we say that the length of a bar is indirectly proportional to the frequency? _____

1.6 What is the main feature of the spectrum of the rectangular bar?

Circular Membranes

2.1 In the figure, below draw several modes which you observed during the demonstration.

Figure 8.1

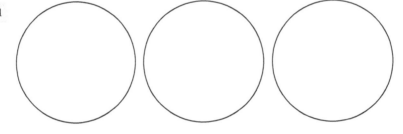

2.2 What kinds of modes do you see in this setup? Do you see any radial modes? _____

2.3 Do the modes you see have a node or antinode at the center? _____

2.4 In the figure below, make a schematic sketch of several radial modes.

Figure 8.2

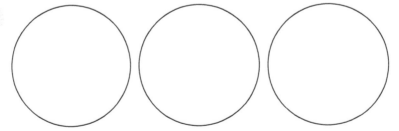

2.5 Does the radial mode have a node or antinode in the center? _____

2.6 What is the main feature of the spectrum of the circular membrane? Do modes of membrane belong to a harmonic series?

Square Membrane

3.1 Draw several modes of a square membrane based on the demonstration.

Figure 8.3

3.2 What changes the picture of the modes: touching the membrane at the nodal line or at the antinode?

3.3 Draw one of the modes where the marked point is fixed (watch the demonstration).

Figure 8.4

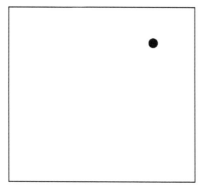

Conclusions

1. What modes of the circular membrane are excited when we hit the membrane exactly at the center?
    a.   Circular modes
    b.   Radial modes
    c.   All modes of the membrane
    d.   No modes at all.

2. A metallic bar of length L has a frequency of the lowest mode, f. You cut this bar into two halves. What will be the lowest frequency of each half?
    a.   2f
    b.   4f
    c.   F
    d.   f/2
    e.   f/4.

NORMAL MODES OF STRINGS: GUITARS, VIOLINS, AND THE STICK-SLIP MECHANISM

Modes of Guitar Strings

1.1 In the figure below, draw the first three modes of a guitar string. Mark the nodes (n) and antinodes (a).

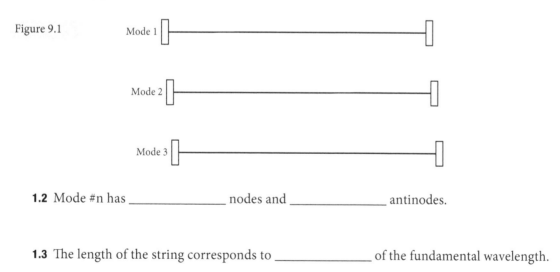

Figure 9.1

Mode 1

Mode 2

Mode 3

1.2 Mode #n has _____ nodes and _____ antinodes.

1.3 The length of the string corresponds to _____ of the fundamental wavelength.

1.4 Write the relationship between the frequency of harmonic #n, speed of wave, and length of string:

1.5 A string vibrates in mode #5. At what fraction of length L is the first node from either end? _____

1.6 A string vibrates in mode #5. At what fraction of length L is the first antinode from either end? _____

1.7 In an upper part of the figure below, schematically draw a string which is vibrating in mode #2. Now, in the lower part of the figure, draw what kind of mode you'll get if you fix the string at L/4 (point is shown).

Figure 9.2   Mode 2

L/4

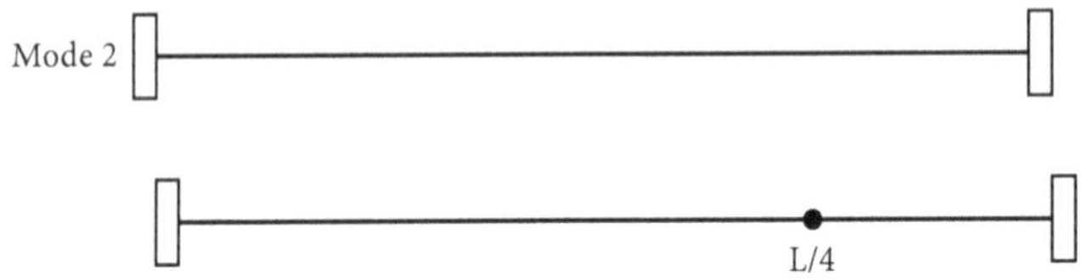

1.8 What first three modes will be left out of the recipe if you pluck the string at L/5?

_____

1.9 To conclude: what is the main feature of the spectrum of a string? Does the string naturally have a harmonic series?

Guitar Maker's Rule

2.1 What is the pitch difference between two adjacent frets on a guitar? _____

2.2 At what fraction of the total length of string L should we position the first fret from a bridge? _____

2.3 How does the distance between frets change while we move closer to the sound hole? _____

2.4 The string in the figure below has a total length of 36 inches. Mark the first three frets and calculate the distance between them.

Figure 9.3 Bridge

Violin

3.1 The figure below shows a cross-section of a violin "cut" at its waist. The bridge is shown and the strings are marked. Draw the sound post and bass bar.

Figure 9.4

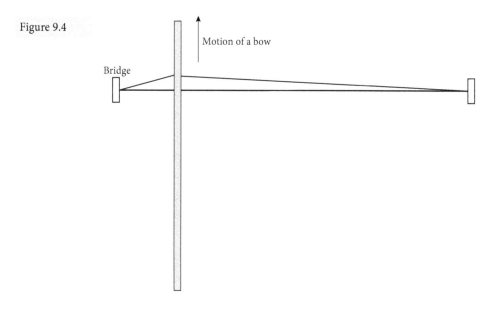

Motion of a bow

Bridge

3.2 The figure below shows a string with a bow. The direction of motion of the bow is also shown. Indicate the direction of motion of a kink after releasing.

Figure 9.5

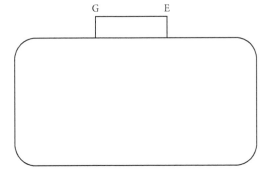

3.3 In your own words, describe the main difference between good violins and generic violins.

Conclusions

1. A string vibrating in mode #7 has:
   a.   6 nodes and 6 antinodes
   b.   7 nodes and 7 antinodes
   c.   6 nodes and 7 antinodes
   d.   7 nodes and 6 antinodes.

2. What modes disappear from the recipe of a sound when we pluck a string at L/7?
   a.   7, 9, 11
   b.   7, 14, 21
   c.   All odd modes
   d.   All even modes.

Activity 10:

FLUTES, RECORDERS, AND REEDS: AIR COLUMN VIBRATIONS AND NORMAL MODES

1. Pipes are open from both ends

1.1 In the figure below, draw the first three displacement modes and mark the nodes (n) and antinodes (a).

Figure 10.1

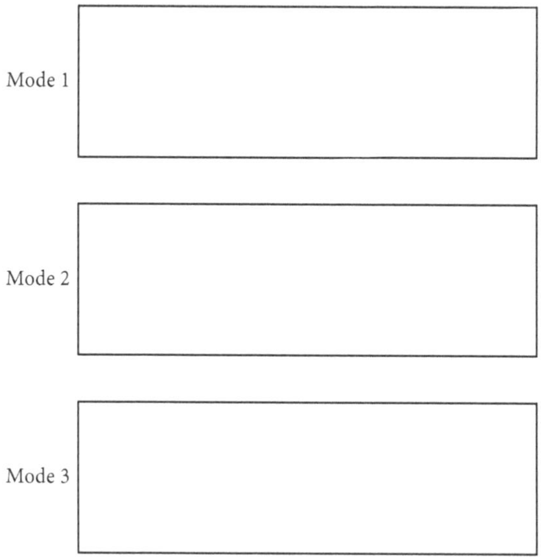
Mode 1
Mode 2
Mode 3

1.2 Mode #n has _____ nodes and _____ antinodes.

1.3 The length of a pipe with both ends open corresponds to _____ of fundamental wavelength.

1.4 Write the relationship between the frequency of harmonic #n, the speed of wave, and the length of pipe:

1.5 Let us consider the change of voice of a pipe with a change of temperature.

1.5.1 What is the fundamental frequency of a pipe of 1 m long at 0°C? (Hint: first calculate the speed of sound in the air.)

1.5.2 What is the fundamental frequency of a pipe of 1 m long at 30°C? (Hint: First calculate the speed of sound in the air.)

1.5.3 What is the ratio of these frequencies? _____

1.5.4 What is the nearest musical interval to this ratio? _____

1.5.5 How does the pitch of a pipe change with the temperature rising? _____

1.6 What is the main feature of a spectrum of a cylindrical pipe? Does it demonstrate a harmonic series?

Fingerholes

2.1 If we drill a large hole in a cylindrical pipe, this hole serves as:

a. Additional displacement node
b. Additional displacement antinode
c. May be a node or an antinode.

2.2 You drill a hole exactly in the center of a cylindrical pipe.

2.2.1 How will the opening of this hole change the effective length of your pipe? _____

2.2.2 How will the opening of this hole change the pitch? What is the ratio of pitch with the hole open to the pitch with the hole closed? _____

2.3 What instruments do you know of that use the pipe with both ends open?

3. A pipe is closed from one end.

3.1 In the figure below, draw the first three displacement modes and mark the nodes (n) and antinodes (a).

Figure 10.2

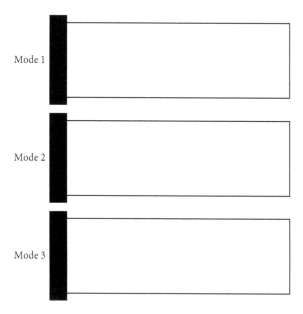

Mode 1

Mode 2

Mode 3

3.2 The length of a pipe with both ends open corresponds to _____ of fundamental wavelength.

3.3 Define the relationship between the frequency of the harmonic #n, speed of wave, and length of pipe:

3.4 Does the pipe with one end closed have both odd and even harmonics? _____

3.5 What instruments do you know of that use the pipe with one end closed?

3.6 To conclude: what happens to the pitch of a pipe if we close one end? _____

Reeds

4.1 In the figure below, show the direction of motion of two sheets of paper when you blow air between them.

Figure 10.3

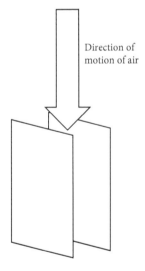

Direction of
motion of air

4.2 What do we call this effect?

4.3 What other examples from everyday life can you name?

Conclusions

1. A pipe may resonate at frequencies of 600, 800, and 1000 Hz, but nothing in between. This pipe is:

    a.   A pipe with both ends open and a fundamental of 400 Hz

    b.   A pipe with both ends open and a fundamental of 200 Hz

    c.   A pipe with one end closed and a fundamental of 200 Hz

    d.   A pipe with one end closed and a fundamental of 400 Hz.

2. What instrument uses a cylindrical pipe closed from one end and a single reed?

    a.   Flute

    b.   Saxophone

    c.   Clarinet

    d.   Oboe.

THE HUMAN VOICE: FORMANTS

General Properties of the Human Voice Apparatus

1.1 The human voice is just another musical instrument. What family of musical instruments (or just a particular musical instrument) does it resemble?

1.2 What is the usual range of a human voice? _____

1.3 What is the usual range of a female voice and a male human voice?

Female _____

Male _____

1.4 Do you know of any singers with a much broader range of voice? (optional)

Vowel Production

2.1 What part of the voice apparatus is responsible for the height of a pitch?

2.2 How do we regulate the height of a pitch? (Hint: By tension of vocal cords or by the configuration of the vocal tract?)

2.3 What part of the voice apparatus is responsible for articulation?

2.4 What are we changing when we say "cat" and then say "caught?"

Formants

3.1 The figure below shows the centers of two formants for a vowel, say, "ee." ("Ee" is only used here as an example.)

Figure 11.1

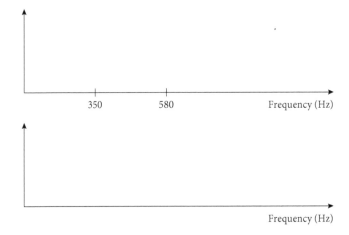

3.1.1 Schematically draw these formants in the upper graph.

3.1.2 What defines the position of these formants? (Hint: tension of vocal cords or the character of a spoken word?)

3.1.3 Now, sing the vowel whose formants correspond to the picture shown (we don't know what this particular vowel is). The pitch of a vowel sung is 100 Hz. Draw the harmonics that are delivered into your vocal tract with this vowel. Draw four to five harmonics and mark their frequencies.

3.1.4 Now, in the lower system of coordinates, mark the formants for the same vowel as in questions 3.1.1–3.1.3, but for "ee" sung one octave higher. Did the positions of the formants change? (Hint: try it yourself: when you say "ee" one octave higher, does it change the configuration of your mouth?)

3.1.5 Draw the harmonics delivered by the vocal cords into your vocal tract as shown in question 3.1.4. Draw four or five of them and mark the frequencies.

3.1.6 To conclude: compare the upper and lower pictures in Figure 11.1. What is changed? Explain in your own words.

Whisper versus Loud Voice

4.1 The figure below shows the centers of two formants for a vowel, say, "oh." ("Oh" is only used here as an example.

Figure 11.2

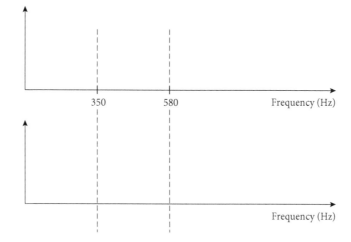

4.1.1 Schematically draw these formants in the upper and lower graphs. Are they identical?

4.1.2 Now, in the upper graph, draw the harmonics delivered to your vocal tract by loudly saying "oh" at a pitch of 100 Hz.

4.1.3 In the lower graph, draw the harmonics delivered to your vocal tract by whispering "oh" at the same pitch of 100 Hz.

4.1.4 Explain the difference between these two graphs.

4.2 Refer to Figure 15.8 in your book. What is the main difference between the articulations of different words?

4.3 Why do sopranos with a very high range change words while singing?

Falsetto versus Full Voice

5.1 Explain the difference between falsetto and full voice. (Hint: think about tension and motion of vocal cords).

Conclusions

1. The positions of the first and second formants of the word "hood," said with a pitch of 200 Hz, are 400 Hz and 1300 Hz, respectively. What positions of the formants for the same word "hood" are said with a pitch of 400 Hz?
    a.   400 Hz and 2600 Hz
    b.   800 Hz and 2600 Hz
    c.   400 Hz and 1300 Hz
    d.   800 Hz and 1300 Hz.

2. What is the main difference between a whisper and a loud voice?
    a.   In a loud voice, the content of the upper harmonics is very poor
    b.   In a whisper, the content of the upper harmonics is very poor
    c.   The configuration of the vocal tract
    d.   There is no difference at all.

ROOM ACOUSTICS AND A MAP OF A SYMPHONIC ORCHESTRA

1. Reverberation time.

1.1 What is reverberation time?

1.2 What are typical reverberation times for classrooms? _____

1.3 What are typical reverberation times for organ halls? _____

1.4 What is the main difference between organ halls in Europe and America?

1.5 What formula do we use to define reverberation time in the room? Explain all symbols in this formula.

1.6 How does reverberation time change if the absorption coefficient of walls increases?

1.7 What is the desirable dependence of reverberation time on frequency in a concert hall? Draw the dependence in the graph below.

Figure 12.1

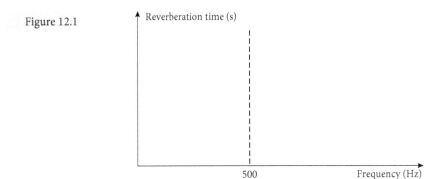

2. Pick a musical instrument.

2.1 In the diagram below, mark the position of the group of your instruments in a symphonic orchestra.

Figure 12.2

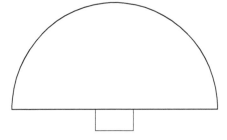

2.2 What defines such a position of your instrument in a symphonic orchestra?

2.3 Describe the properties of your instrument: to what family does it belong, what is the source of the sound, and what are the main features of sound production and spectrum? For this, you may use your book or just do an Internet search with your cellphone.

2.4 How does the change in temperature affect the pitch of your instrument?

1. With an increase of frequency, the reverberation time of a typical room

    a.   Increases

    b.   Decreases

    c.   Stays the same

    d.   Impossible to predict.

2. Typical reverberation times of classrooms are:

    a.   0.5–1 s

    b.   5 s

    c.   12 s

    d.   Could be any.

CPSIA information can be obtained
at www.ICGtesting.com
Printed in the USA
BVHW050650220122
626775BV00006B/502